YOUR KNOWLEDGE HAS VALUE

- We will publish your bachelor's and
  master's thesis, essays and papers

- Your own eBook and book -
  sold worldwide in all relevant shops

- Earn money with each sale

Upload your text at www.GRIN.com
and publish for free

Urgessa Tilahun

Challenges and prospects of agricultural production and productivity

GRIN Verlag

Bibliografische Information der Deutschen Nationalbibliothek:

Die Deutsche Bibliothek verzeichnet diese Publikation in der Deutschen National-
bibliografie; detaillierte bibliografische Daten sind im Internet über http://dnb.d-
nb.de/ abrufbar.

Imprint:

Copyright © 2013 GRIN Verlag GmbH
Druck und Bindung: Books on Demand GmbH, Norderstedt Germany
ISBN: 978-3-656-72402-5

This book at GRIN:

http://www.grin.com/en/e-book/279601/challenges-and-prospects-of-agricultural-
production-and-productivity

CHALLENGES AND PROSPECTS OF AGRICULTURAL PRODUCTION AND PRODUCTIVITY IN ETHIOPIA

Urgessa Tilahun Bekabil

Haro Sabu Agricultural Research Center, Haro Sabu, Ethiopia

I. TABLE OF CONTENTS

II.   ACKNOWLEDGEMENT

This seminar would never be completed without the contribution of many people to whom I would like to express my thanks. I wish to thank my best friend Emiru Dirbaba and my former staffs of East Wollega Zone Finance and Economic Development Office. Above all, I would like to praise my God for his unspeakable gift in all my life in general and his help in the process of undertaking this study in particular.

III.   LISTS OF ABBREVIATIONS

ADLI	Agricultural Development Led Industrialization
AfDB	African Development Bank
CSA	Central Statistical Agency
FAO	Food and Agricultural Organization
FISIM	Financial Intermediary Service Indirect Measurement
GDP	Gross Domestic Product
GNP	Gross National Product
IFPRI	International Food Policy Research Institute
MoARD	Ministry of Agriculture and Rural Development
MoFED	Ministry of Finance and Economic Development
NBE	National Bank of Ethiopia
PADETES	Participatory Demonstration and Training Extension System
R&D	Research and Development
SSA	Sub Saharan Africa
UNDP	United Nations Development Programme
UoP	University of Pretoria
USD	United States Dollar

V.   LIST OF FIGURES

VI.   ABSTRACT

Agricultural production in Ethiopia is characterized by subsistence orientation, low productivity, low level of technology and inputs, lack of infrastructures and market institutions, and extremely vulnerable to rainfall variability. Productivity performance in the agriculture sector is critical to improvement in overall economic well-being in Ethiopia. Low availability of improved or hybrid seed, lack of seed multiplication capacity, low profitability and efficiency of fertilizer, lack of irrigation development, lack of transport infrastructure, inaccessibility of market and prevalence of land degradation, unfertile soil, overgrazing, deforestation and desertification are among the constraints to agricultural productivity during last period. However, in 2011 the sector grew by 9% driven by cereal production which reached a record high of 19.10 million tons in Ethiopia.

1. INTRODUCTION

The history of agricultural and rural development since the end of World War II in 1945 is characterized by changing priorities and concerns. Immediately after this war and the widespread experience of serious malnutrition, there was a determined effort to increase food production in the developed world (Reimund *et al.*, 2007).

Rural areas are home to 75 percent of Africa's people, most of whom count agriculture as their major source of income. Fortunately, Africa has experienced continuous agricultural growth during the past few years. However, much of the growth has emanated from area expansion rather than increases in land productivity. In most countries, future sustainable agricultural growth will require a greater emphasis on productivity growth, as suitable area for new cultivation declines, particularly given growing concerns about deforestation and climate change (IFPRI, 2012).

Sub-Saharan Africa is one of the world's poorest regions. Its population and land area are approximately three times that of the USA. The region's economies are heavily dependent on agriculture, which accounts for two-thirds of the labour force, 35% of GNP and 40% of foreign exchange earnings. Productivity performance in the agricultural sector is thus critical to improvement in overall economic well-being in Sub-Saharan Africa (Lilyan *et al.*, 2004).

According to Paul *et al.* (2012) agricultural production and proximity (as measured by travel time) to urban markets are highly correlated in Sub-Saharan Africa, even after taking agro ecology into account. According to IFPRI (2012) poor resource endowments, minimal use of inputs (fertilizer, improved seeds, and irrigation), and adverse policies that continued for a long period have been identified as the major causes of the low and declining performance of the agricultural sector in SSA. Continuing environmental degradation, high population growth, and low levels of investment in agricultural infrastructure also aggravate the resource limitations of agriculture in Africa.

Key constraints to agricultural productivity in Ethiopia include low availability of improved or hybrid seed, lack of seed multiplication capacity, low profitability and efficiency of fertilizer use due to the lack of complimentary improved practices and seed, and lack of irrigation and water constraints. In addition, lack of transport infrastructure and market access decreases the profitability of adopting improved practices (Kate & Leigh, 2010).

According to Mulat *et al.* (2004) the Ethiopian economy is among the most vulnerable in Sub-Saharan Africa. It is heavily dependent on agricultural sector (Berhanu, 2009), which has suffered from recurrent droughts and extreme fluctuations of output. As explained by Jordan *et al.* (2011) the opportunities and constraints facing Ethiopian agriculture are strongly influenced by conditions which vary across geographical space. These conditions include basic agricultural production potentials, access to input and output markets, and local population densities which represent both labor availability and local demand for food.

The impact of rainfall on crop production can be related to its total seasonal amount or its intra-seasonal distribution. In the extreme case of droughts, with very low total seasonal amounts, crop production suffers the most. But more subtle intra-seasonal variations in rainfall distribution during crop growing periods, without a change in total seasonal amount, can also cause substantial reductions in yields. This means that the number of rainy days during the growing period is as important, if not more, as that of the seasonal total. Generally, the effect of rainfall variability on crop production varies with types of crops cultivated, types and properties of soils and climatic conditions of a given area (Woldeamlak, 2009).

The agricultural sector, which accounts for 80% of employment, remains a key source of growth. In 2011 the sector grew by 9%, driven by cereal production which reached a record high of 19.1 million tons in 2011. Agricultural production has been boosted by favorable weather conditions in cereal-growing areas, enhanced government support services to smallholders, improvement in yields and expansion in the area under cultivation. Increases in productivity are mainly responsible for increased yields, rather than extension of the cultivated area. This is consistent with the government's massive push to promote and deliver technology packages to smallholders (AfDB, 2012).

Increasing productivity in smallholder agriculture is the Government's top priority. This recognizes that: (i) smallholder agriculture is the most important sub-sector of Ethiopia's economy; (ii) there remains a high prevalence of poverty among smallholder farming communities; and (iii) there is a large potential to improve crop and livestock productivity using proven, affordable and sustainable technologies (MoARD, 2010).

2. OBJECTIVE OF THE SEMINAR

The objective of this paper was the following;

- to describe the prospects of agricultural production and productivity in Ethiopia

- to assess the challenges of agricultural production and productivity during implementation of different agricultural policies and strategies including the Growth and Transformation Plan (GTP) and

- to suggest policy measures that could minimize the challenges of agricultural production and productivity in order to attain the planned activities thereon Growth and Transformation Plan (GTP).

3. REVIEW OF LITERATURE

3.1. Agricultural Production and Productivity in Sub-Saharan Africa

As stated by Lilyan *et al.* (2004) the Sub-Saharan Africa countries showed some progress in the 1960s, suffered a regression in productivity during the 1970s, but after the mid-1980s recovered to achieve a reasonably robust rate of productivity improvement through the end of the century. The overall average rate of productivity growth for the four decades was estimated at 0.8% per year.

As stated by Paul *et al.* (2012) the impacts of investments in road infrastructure on agricultural output and productivity are particularly important in Sub-Saharan Africa for three reasons. First, the agricultural sector accounts for a large share of gross domestic product (GDP) in most Sub-Saharan countries. Second, poverty is concentrated in rural areas. Finally, the relatively low levels of road infrastructure and long average travel time result in high transaction costs for sales of agricultural inputs and outputs, and this limits agricultural productivity and growth. Thus, investments in road infrastructure and related transport services can have a significant impact on rural and national incomes through their effects on agriculture (Paul *et al.*, 2012).

3.2. Agricultural Production System in Ethiopia

Agricultural production is dominated by smallholder households which produce more than 90% of agricultural output and cultivate more than 90% of the total cropped land. Smallholders drive their income either in cash or through own-consumption from agricultural production. According to the national accounts, the agricultural sector consists of crop, livestock, fishery and forestry sub-sectors. Crop production is the dominant sub-sector within agriculture, accounting for more than 60% of the agricultural GDP followed by livestock which contributes more than 20% of the agricultural GDP. The contributions of forestry, hunting and fishing do not exceed 10% (Mulat *et al.*, 2004).

The viability of the agricultural production systems in Ethiopia, as in many areas in developing countries, is highly constrained by degraded soils and increasing lack of reliability in rainfall resulting from climate change (Menale *et al.*, 2010). There are two main production systems in Ethiopia: the pastoral nomadic system, and the mixed crop production system. The pastoral livestock production system dominates the semi-arid and arid lowlands (usually below 1500 meters above sea level). These regions cover a vast area of lands with a small livestock production.

The crop production system can be classified into smallholders' mixed farming, producers' cooperative farms, state farms, and private commercial farms based on their organizational structure, size, and ownership. The major objectives of small holder farmers' production are to secure food for home consumption and to generate cash to meet household needs such as clothing, farm inputs, taxes and others.

Ethiopia has a variety of fruits, leafy vegetables, roots and tubers adaptable to specific locations and altitudes. The major producers of horticultural crops are small scale farmers, production being mainly rain fed and few under irrigation. Shallot, garlic, potatoes and chillies are mainly produced under rain fed conditions. Tomatoes, carrots, lettuce, beetroot, cabbage, spinach and swiss chard are usually restricted to areas where irrigation water is available (Girma, 2003).

Ethiopia has got an immense potential to develop intensive horticulture on small scale as well as on commercial scale. According to Girma (2003), some of the favorable factors that contribute to an overall investment are:

- Proximity to lucrative markets,
- Agro-climatic suitability and rich water resources for diversified irrigated agriculture,
- Growth/rise of demand for horticultural crops, particularly in urban areas,
- Diversified agro-climatic conditions that facilitate the diversification of the crops,
- The high productivity of horticultural crops as compared to cereals,
- Export possibilities of these crops are very encouraging and
- If fully exploited, these crops are highly remunerative and would be undoubtedly help to improve the standard of living of small scale resources poor farmers.

3.3. Challenges of Agricultural Production and Productivity in Ethiopia

The agricultural production trends throughout the 1980's up to mid-1990's were characterized by wide fluctuations in total output and weak growth, with grain production increasing at rate of 1.37% annually compared to population growth of 2.9 % (World Bank, 2004).

Despite large scale extension efforts since mid-1990s, agricultural performance over the past decade has continued to be weak, with production gains mainly driven by weather and area expansion, and weak yield gains limited to maize. There are multiplicity of factors explaining this poor performance; high rainfall variability, the lack of irrigation investment, weak rural institution limited modern varieties, the lack of animal traction, the lack of mechanization, under investment in agricultural research, weak rural infrastructure and skills on the demand side, poor market linkages, high transaction costs, and weak purchasing power leading to thin and volatile markets, make agriculture more risky and reduce production incentives (World Bank, 2004).

Investments in productivity increases higher up the food value chain, such as through marketing and transportation infrastructure, would increase prices farmers receive for output while also putting downward pressure on urban food prices. Higher producer prices would create incentives for farmers to invest in productivity increasing technologies since output increases would offer substantial gains (Kate & Leigh, 2010).

Because of the diverse agro-ecological zones, topography and natural vegetations, Ethiopian small farmers have developed complex farming methods and cropping patterns. Accordingly, seven different cereal crops, six pulse crops, seven oilseed crops, and a number of different other and tree crops are grown. Diversification has allowed farmers to cope with the drought or erratic rains but identifying the right technological package for the various ecologies and crops has been of considerable challenge to researchers and extension systems (Mulat *et al.*, 2004).

According to Mulat *et al.*, (2004), Ethiopia is said to possess the largest livestock population in Africa. Livestock is considered as a security during crop failure, investment and additional income for farmers in Ethiopia. Livestock serve as source of traction for crop production, raw material input for industry (e.g. hides and skins, wool, etc.) and manure for fertilization. Equines are the major source of transport services in rural areas. The role of livestock as a source of food is critical for both highland and lowland inhabitants.

The main food contributions of livestock include, among other things, meat and meat products, milk and milk products, eggs, and honey. In mixed farming systems of the highlands, 26% of the livestock output is used as food, while in the pastoral areas, where livestock forms the main sources of livelihood, this proportion increases to 61%.

Despite its potential, the livestock sub-sector has remained undeveloped in Ethiopia. On average it contributes up to 30 percent of agricultural GDP. The main constraints (Mulat *et al.*, 2004) include the following:

Diseases: Diseases have been identified one of the main factor for low productivity of the livestock sub-sector. About 30-50% of the total value of livestock products is lost every year due to diseases such as rinderpest, trypanosomiasis, foot and mouth disease, and liver fluke.

Feed shortage: under-nutrition and malnutrition are among the major constraints of livestock production in Ethiopia. Nutritional stress has caused low growth rates, poor fertility and high mortality. High population growth and increasing density have led to expansion of cultivated area at the cost of grazing land on which smallholder livestock production depends. Permanent pastureland is believed to have declined by close to 60% over the last three decades. It should be note that in areas where there is intensive cultivation, crop residues have become the main source of animal feed.

Demand constraint: underdevelopment of roads and other infrastructure has hindered livestock take-off. It has been indicated that as income declines for a variety of reasons, livestock products are the first to be selected or removed from the menu by the majority of consumers. Also, during fasting seasons (which are many) of Christians, livestock products are not part of the daily menu, i.e. they are not entirely consumed which influences the demand for products negatively.

Institutional and policy constraints: there are also institutional and policy related problems such as lack of institutional stability that could promote the sub-sector, lack of appropriate policies to promote and increase production and productivity of the sub-sector. Inadequate capital and recurrent budget allocations to the livestock sub-sector have also contributed to its low productivity.

The agricultural sector continues to face major challenges. Rural livelihoods remain extremely vulnerable to meteorological shocks, as food production is mainly rain-fed. Despite improvements, productivity levels are still very low and the marketing infrastructure is also weak, leading to high transaction costs. The limited use of improved farming practices by the majority of smallholders is an important factor contributing to low productivity (AfDB, 2012).

The rising cost of key agricultural inputs (e.g. Chemical fertilizer) is another challenge in promoting modern technology packages to improve productivity. Furthermore, soil erosion in certain agro ecological zones due to over-cultivation and limited investment in land improvement is a hindrance to sustainable agricultural output growth. There has been a general decline in per capita food production as high population growth rates have contributed to a decline in farm size. Hence about 4.5 million people remain dependent on food relief (AfDB, 2012).

Soil erosion is one of the major agricultural problems in the highlands of Ethiopia. Deforestation, overgrazing, and cultivation of slopes not suited to agriculture together with the farming practice that do not include conservation measures are the major causes for soil erosion in much of Ethiopia's highland areas. Degraded soils are also the major constraints to agricultural production and food security in the Southern Ethiopian highlands (Abay, 2011).

Land degradation is one of the major causes of low and in many places declining agricultural productivity and continuing food insecurity and rural poverty in Ethiopia. As stated by Berry (2002) the major interacting root causes of land degradation in Ethiopia are the following; the impact of natural conditions especially periodic drought, inaccessibility of rural areas due to topographic constraints, steady growth of population and livestock totals without changes in agricultural and other economic systems, historical patterns of feudal ownership of land followed by government ownership and despite policy changes uncertain status of land ownership, institutional overlap, duplication of effort and shortage of financial resources, lack of rural infrastructure and markets, lack of participation of stakeholders in management decisions especially at the local level, weak extension services and low technology agriculture, leading to risk aversion and reliance on cattle as wealth.

The direct causes of land degradation are mainly deforestation, overgrazing and over-cutting, shifting cultivation and agricultural mismanagement of soil and water resources: such as non-adoption of soil and water conservation practices, improper crop rotation, use of marginal land, insufficient and/or excessive use of fertilizers, mismanagement of irrigation schemes and over pumping of groundwater. The indirect causes of land degradation are mainly population increase, land shortage, short-term or insecure land tenure and poverty and economic pressure (FAO, 2001).

Soils in most SSA countries have inherent low fertility and do not receive adequate nutrient replenishment. Soil productivity in SSA is also constrained by aridity (low rainfall) and acidity. Although little production increase has taken place, this has been obtained by cultivation of poor and marginal lands while the productivity of most existing lands has been declining (FAO, 2001).

According to Abay (2011) soil erosion is one of the major agricultural problems in the highlands of Ethiopia. Deforestation, overgrazing, and cultivation of slopes not suited to agriculture together with the farming practice that do not include conservation measures are the major causes for soil erosion in much of Ethiopia's highland areas. Degraded soils are also the major constraints to agricultural production and food security in the Southern Ethiopian highlands.
Low and declining soil fertility is one of the major causes of poor yields and the loss of fertile topsoil as a result of erosion and desertification has seriously reduced the production potential of previously fertile lands. Opportunities to raise yields and increase land and labour productivity through improved soil management and water conservation rely heavily on the use of external (yield-increasing) inputs (Reimund *et al.*, 2007).

Animals are a major part of the food production system in the arid, semi-arid and sub humid regions. The value of animal manure in crop production has long been widely recognized, and is essential for sustainable crop production in most low and intermediate input systems. If cattle population increases without restriction, the pressure on grazing areas leads to a loss of edible vegetation and dominance of shrub species, and further to desertification (FAO, 2001).

Deforestation and desertification are both causes and symptoms of the deteriorating productivity as the problems are cyclical: intensive land use - induced by the need to produce more - causes soil degradation and lower yields, which demand that more land is cleared for cultivation (FAO, 2001). Deforestation is indiscriminate cutting or over-harvesting of trees for lumber or pulp, or to clear the land for agriculture, ranching, construction, or other human activities. Deforestation in Ethiopia is due to locals clearing forests for their personal needs, such as for fuel, hunting, agriculture, and at times for religious reasons. The main causes of deforestation in Ethiopia are shifting agriculture, livestock production and fuel in drier areas. Desertification is found on every continent except Antarctica, but international attention has focused mostly on Africa, particularly the region known as the Sahel, the region of northern Africa immediately to the south of the Sahara desert.

According to Samuel (2003) due to the subsistence-oriented mode of production, limited use of purchased farm inputs, unreliable weather conditions, low output prices and uncertainties associated with policies, Ethiopian peasants have little interest for agricultural credit. The only exception is fertilizer credit. This is partly due to high government interest to expand the use of fertilizer that is considered by Ethiopian policy makers as a strategic input to increase cereals production, and partly due to farmers' interest to counterbalance decline in soil productivity that resulted from continuous mono-cropping practices of Ethiopian peasants. Even the fertilizer credit market has faced many problems that have serious implications on the sustainability of the market and farmers' interest for fertilizers.

Production in the rural non-farm sector is highly elastic. That is because there is normally a large supply of underemployed labor, or because labor productivity at very low levels of productivity can be increased with little or no investment. That contrasts with agriculture, which because of the land constraint is inelastic in its supply. In agriculture, production is increased by technological change which shifts the production function. The demand for agricultural output is highly elastic because agricultural goods are tradable including on international markets. Of course, other sources of rural income increase may have a similar multiplier effect on the rural non-farm sector. They are however all very small compared to agriculture (John and Paul, 2010).

Numerous constraints to yield and productivity growth have been identified, including relatively low levels of input use (fertiliser, pesticide, improved seeds), low levels of irrigation, soil degradation and soil erosion, inadequate agricultural research and extension, and constraints in market development (Alemayehu *et al.*, 2011).

3.4. Prospects of Agricultural Production and Productivity

As stated by MoARD (2010) increasing productivity in smallholder agriculture is the Government's top priority. This recognizes that: (i) smallholder agriculture is the most important sub-sector of Ethiopia's economy; (ii) there remains a high prevalence of poverty among smallholder farming communities; and (iii) there is a large potential to improve crop and livestock productivity using proven, affordable and sustainable technologies.

The contribution of agriculture to food security both through its direct impact on food production and indirect effect on farm incomes (i.e. through improving entitlement capacity) has failed to recover even after the economic reforms of the 1990s. Despite some short-lived successes in some areas and years, the impact of the country's new development strategy that is commonly known as ADLI and its main instrument, PADETES (the agricultural extension system that was designed based on ADLI strategy) was too little to affect per capita agricultural production or productivity at national level or in a sustainable manner (Samuel, 2003).

In 2011, Ethiopia was ranked 174th out of 187 countries in the UNDP human development index, with a GDP per capita adjusted with the Purchasing Power Parity of USD 971 (compared to almost USD 2 000 average for Sub-Saharan countries). After a significant contraction in 2003/04 due to a severe drought that affected agricultural production, the Ethiopian economy has experienced a broad-based and steady growth of real GDP. In general, the main determinants of the sustained economic growth are the good performance of agricultural production, with significant contribution of manufacturing and services as well as the expansion of the construction sector (mainly housing, roads and hydroelectric dams).

Ethiopia continued to maintain the double digit growth rate which averaged 11.4 percent over the last eight years. In the fiscal year 2010/11, real GDP growth was 11.4 percent moderately higher than the previous year's growth of 10.4 percent (NBE, 2011). Regarding sectoral development, agriculture grew by 9 percent, industry 15 percent and services 12.5 percent. Consequently, agriculture and allied activities accounted for 41 percent of GDP, industry 13.4 and services 45.6 percent. Similarly, agriculture contributed 4.7, industry 1.5 and service 5.3 percentage points to the 11.4 percent real GDP growth in 2010/11. Although, the share of agriculture in GDP tended to decline over time, it still remains the largest employer, the main source of foreign exchange, and supplier of raw materials and market to domestic industries

Table 1 Sectoral Contribution to GDP and GDP Growth (In Millions of Birr) and Share in GDP (in percent)

Sectors	Sectoral Contribution to GDP and GDP Growth (In Millions of Birr)						
	2004/05	2005/06	2006/07	2007/08	2008/09	2009/10	2010/11
Agriculture	39,728	44,062	48,225	51,843	55,141	59,348	64,698
Industry	11,402	12,561	13,757	15,150	16,616	18,374	21,178
Service	33,312	37,747	43,534	50,519	57,576	65,084	73,368
Total	84,442	94,370	105,516	117,512	129,333	142,806	159,244
Less FISM	639	896	1,018	1,323	1,489	1,619	1780
Real GDP	83,804	93,474	104,499	116,190	127,844	141,187	157,464
Sectors	Share in GDP (in percent)						
	2004/05	2005/06	2006/07	2007/08	2008/09	2009/10	2010/11
Agriculture	47.4	47.1	46.1	44.6	43.1	42.0	41.1
Industry	13.6	13.4	13.2	13.0	13.0	13.0	13.4
Service	39.7	40.4	41.7	43.5	45.0	46.1	46.1

Source: MoFED

Figure 1 Sectoral contribution to in millions of birr

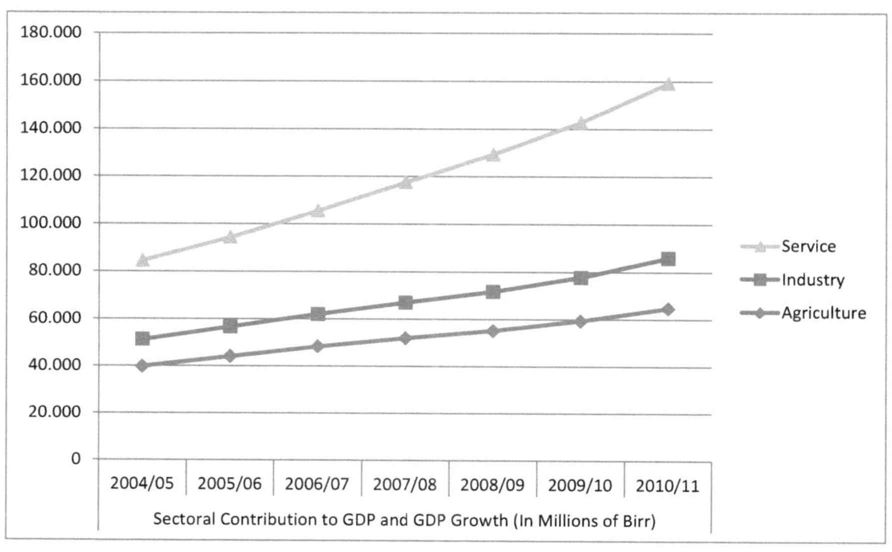

Source: Computed from table 1

Figure 2 Share in GDP in percent

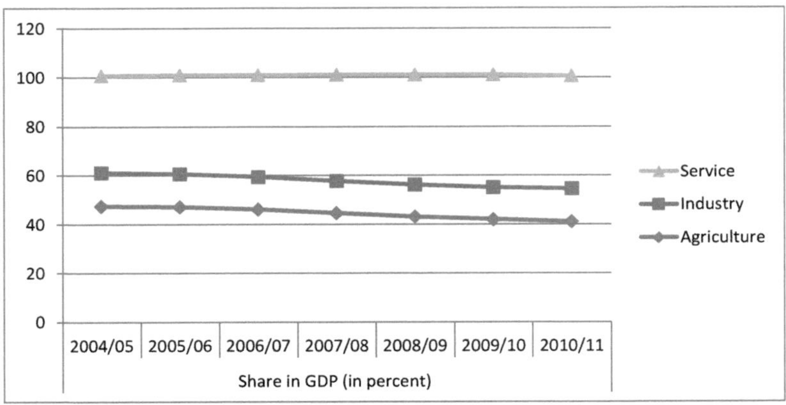

Source: Computed from table 1

The growth in agricultural outputs was largely attributed to improved productivity aided by favorable weather condition and conducive economic policy. Cultivated land expanded by 4.6 percent and reached 12 million hectares in 2010/11. Production is estimated to have increased by about 8.8 percent while productivity rose from 15.7 quintal/hectare in 2004/05 to 16.3 quintal per hectare in 2010/11 Cereal production accounted for about 87.7 percent of the total production estimated for 2010/11(NBE, 2011).

Meanwhile, the 15 percent annual growth in industry was largely due to expansion in electricity and water subsectors. Manufacturing grew by 12 percent with mining and quarrying expanded by 57.7 percent. The 12.5 percent growth in service sector which has gained momentum in recent years was attributed to growth in financial sector, real estate and hotel and tourism sectors (NBE, 2011).

Table 2 Estimates of agricultural production and cultivated areas of major crops for private Peasant Holdings during Meher season

Agricultural Production	2007/08		2008/09		2009/10		2010/11	
	Area Cultivated	Total Production	Area Cultivated	Total Production	Area Cultivated	Total Production	Area Cultivated	Total Production
Cereals	8,730.0	137,169.9	8,770.0	144,964.1	9,233.0	155,342.2	9,905.5	172,383.2
(Percent Change)	3.0	6.5	0.5	5.7	5.3	7.2	7.3	11.0
Pulses	1,517.7	17,827.4	1,585.2	19,646.3	1489.3	18980.5	1343.9	17,487.7
(Percent Change)	10.1	12.9	4.4	10.2	-6.1	-3.4	-9.8	-7.9
Oilseeds	707.6	6169.3	855.1	6,557.0	780.9	6436.1	781.2	6765.2
(Percent Change)	-4.6	24.1	20.8	6.3	-8.7	-1.8	0.04	5.1
Total	10,955.3	161,166.6	11,210.3	171,167	11,503.2	180758.8	12,030.6	196,636.1
(Percent Change)	3.4	7.8	2.3	6.2	2.6	5.6	4.6	8.8

Source: Central Statistical Agency

Notwithstanding the constraints, the potential for growth in agriculture is huge, especially considering that less than 15% of the arable land is cultivated while productivity is still among the lowest in sub-Saharan Africa. Given these low initial conditions significant expansion in agricultural production within a short period of time is possible by scaling up the diffusion of improved farming practices, deepening commercialization and improving the security of land tenure.

Much of the increase in crop production in the past decade has been due to increases in area cultivated. To what extent the area cultivated can continue to expand remains an important question. It seems that in the highland areas, expansion of cultivated area will have to come almost exclusively from reduction in pasture land. In most instances, this land is likely to be less fertile than existing crop land. Increased use of inter-cropping or double cropping may allow some expansion of area cultivated as well. Expansion of area cultivated outside of the highland regions will require major investments in infrastructure and might involve reductions in forest areas, with important negative environmental implications. As a consequence, it seems that obtaining higher yield rates is *the* challenge of Ethiopia's agricultural system (Alemayehu *et al.*, 2011).

Building on successful experience with model farmers, the GTP seeks to transform agriculture by providing incentives for the commercialization of agriculture while continuing to support smallholders to raise productivity. Thanks to the incentives that the government is providing, foreign investment in commercial farming is expanding, which is expected to boost food production and exports of commercial crops. Weather conditions remain a critical factor but prospects for the agricultural sector to perform w ell in 2012 and 2013 are favorable in view of the recent strides in commercialization and the positive response of smallholder farmers to support services (AfDB, 2012).

4. CONCLUSION AND RECCOMENDATION

4.1. Conclusion

Agricultural production in Ethiopia is characterized by subsistence orientation, low productivity, low level of technology and inputs, lack of infrastructures and market institutions, and extremely vulnerable to rainfall variability. The economy of Ethiopia is based largely on low productive rain-fed agriculture where production heavily depends on rain for its success or failure. Ethiopia is endowed with diverse terrain and agro-ecological climate ranging from temperate in the highlands to tropical in the lowlands. Because of this, a variety of crops can be grown in the country. Much of the land area, however, is mountainous. The mountain range and the rift valley run north-south through the center of the country

During the 1990s, most of the increase in cereal production came from increases in area. However, in the 2000s, area and yield increases each accounted for about half of production growth and we thus see an initial start of increasing intensification. With little suitable land available for expansion of crop cultivation available, especially in the highlands, future cereal production growth will need to come increasingly from yield improvements. Current improved inputs use is relatively low, suggesting substantial scope for raising productivity through the increasing adoption of improved seeds and chemical and organic fertilizer, i.e. at least in the rainfall sufficient ecologies (and on irrigated farms). However, it seems that growth in agricultural real incomes will also require more diversification and a shift to higher-value crops, as to respond to changing consumption baskets driven by the increasing per capita income growth in the country.

Key constraints to agricultural productivity in Ethiopia include low availability of improved or hybrid seed, lack of seed multiplication capacity, low profitability and efficiency of fertilizer use due to the lack of complimentary improved practices and seed, and lack of irrigation and water constraints. In addition, lack of transport infrastructure and market access decreases the profitability of adopting improved practices.

The Ethiopian economy is among the most vulnerable in Sub-Saharan Africa. It is heavily dependent on agricultural sector, which has suffered from recurrent droughts and extreme fluctuations of output. The opportunities and constraints facing Ethiopian agriculture are strongly influenced by conditions which vary across geographical space. These conditions include basic agricultural production potentials, access to input and output markets, and local population densities which represent both labor availability and local demand for food.

Increasing productivity in smallholder agriculture is the Government's top priority. This recognizes that smallholder agriculture is the most important sub-sector of Ethiopia's economy and there remains a high prevalence of poverty among smallholder farming communities In addition to the above there is a large potential to improve crop and livestock productivity using proven, affordable and sustainable technologies

4.2.Recommendation

Above all, to increase and improve agricultural production and productivity and to empower farmers in Ethiopia, the key points that should be taken into account are creating conducive political and policy environment, setting up good agricultural system, proper planning and management of resources, sustained government public investments in agricultural technology, and extension irrigation and market infrastructure, creating favorable linkages between farmers, governments, researchers and other agents of agriculture and farmers (whether small of large holder) should be linked to the market, and innovations should address the needs for both group.

5. REFERENCE

Abay, A. 2011. Construction of Soil Conservation Structures for improvement of crops and soil productivity in the Southern Ethiopia, *Journal of Environment and Earth Science*, **1**(1), 2011

AfDB. 2012. African Economic Outlook, Ethiopia, 2012. www.africaneconomicoutlook.org.

Alemayehu, S., Paul, D., and Sinafikeh, A. 2011. Crop Production in Ethiopia: Regional Patterns and Trends, Development Strategy and Governance Division, International Food Policy Research Institute, Ethiopia Strategy Support Program II, Ethiopia

Berhane, H. 2009. The Impact of Agricultural Policies on Smallholder Innovation Capacities, The case of household level irrigation development in two communities of Kilte Awlaelo woreda, Tigray regional state, Ethiopia

Berry. 2002. Land Degradation in Ethiopia: Its Extent and Impact, Commissioned by Global Mechanism with support from the World Bank

FAO. 2001. Soil fertility management in support of food security in sub-Saharan Africa, Food and Agriculture Organization of the United Nations, Rome, 2001

Girma, A.J. 2003 Horticultural Crops Production in Ethiopia, Oromiya Agricultural Research Institute

IFPRI. 2012. Increasing Agricultural Productivity & Enhancing Food Security in Africa, New Challenges & Opportunities, Washington DC, March 2012

John, W. M and Paul, D. 2010. Agriculture and the Economic Transformation of Ethiopia, Development Strategy and Governance Division, International Food Policy Research Institute – Ethiopia Strategy Support Program 2, Ethiopia

Jordan, C., and Emily, S. 2011. Ethiopian Agriculture: A Dynamic Geographic Perspective, Development Strategy and Governance Division, International Food Policy Research Institute – Ethiopia Strategy Support Program II, Ethiopia

Kate, S. & Leigh, A. 2010. Yield Gap and Productivity Potential in Ethiopian Agriculture: Staple Grains & Pulses, University of Washington.

Lilyan, E. F., Richard, K. P., and Bingxin, Y. 2004. Institutions and agricultural productivity in Sub-Saharan Africa, *Journal of Agricultural Economics* **31** (2004) 169–180

Menale, K., Precious, Z., Kebede, M, and Sue, E. 2010. Adoption of organic farming technologies: Evidence from a semi-arid region in Ethiopia, Food and Agriculture Organization of the United Nations, 2010

MoARD. 2010. Ethiopia's Agricultural Sector Policy and Investment Framework (PIF) 2010-2020, 15 September 2010.

Mulat, D., Fantu, G. and Tadele, F. 2004. Agricultural Development in Ethiopia: are there alternatives to food aid?, Addis Ababa University, Addis Ababa, Ethiopia.

NBE. 2011. Annual Report / VII, Investment, 2011

Paul, D., Hyoung, G. W., Liangzhi, Y., and Emily, S. 2012. Road connectivity, population, and crop production in Sub-Saharan Africa, *Journal of Agricultural Economics* **43** (2012) 89–103

Reimund, R., Herman, V.K., Marijke, K., Jan, V. and Gon, V.L. 2007. Science for Agriculture and Rural Development in Low-income Countries, Springer (2007) 1–6

Samuel, G.S. 2003. Summary Report on Recent Economic and Agricultural Policy, Policy Module Ethiopia, International Conference, 20-22, Rome, Italy

Woldeamlak, B. 2009 Rainfall variability and crop production in Ethiopia: Case study in the Amhara region, Department of Geography & Environmental Studies, Addis Ababa University

World Bank. 2004. Can Agriculture Lead Growth in Ethiopia? The Importance of Linkages, Markets and Tradeability